# THE AI SURVIVAL GUIDE FOR STUDENTS

The AI Survival Guide for Students

*What Every Gen Z Should Know About AI to Stay Ahead of The Game*

Nishant Tomar



Published by Game Changer Publishing

Paperback ISBN: 979-8-90158-232-9
Hardcover ISBN: 979-8-90158-233-6
Digital ISBN: 979-8-90158-234-3

GAME CHANGER PUBLISHING

www.GameChangerPublishing.com

# DEDICATION

To every student growing up in a world where algorithms shape what you see, what you believe, and what you become—may you learn to think for yourself, question everything, and stay human.

To my daughters, who inspired me to see AI through the eyes of the next generation.

To my wife, whose love and support have been my constant source of strength; And to my parents, whose sacrifices, values, and belief in me shaped my journey.

# You found the secret bonus level

Scan the QR code to unlock free AI survival resources, student-friendly AI prompts, cheat sheets, and bonus tools that didn't fit inside the book.

**Scan the QR Code here:**

# THE AI SURVIVAL GUIDE FOR STUDENTS

[ What Every Gen Z Should Know About AI to Stay Ahead of The Game ]

Nishant Tomar

# FOREWORD

The world today is abuzz with transformative changes taking place with the proliferation of AI and its applications in many aspects of our lives. There is a wide spectrum of understanding of AI's potential, current uses, ongoing advancements, and the lurking pitfalls across different segments of the population. However, the inevitability of people's lives currently being touched and the future impacts make it absolutely essential for everyone to have a decent understanding of this powerful tool. Nishant Tomar has, in a very timely fashion, come up with an excellent manuscript on a wide range of topics associated with current and future usage of AI, the impact it is currently having, and how it may transform our lives in the future, especially focused on youth and students for whom it is imperative to engage early in this rapidly evolving technology. Nishant himself has experienced his own transformation at work as he has engaged AI tools in two successful businesses he has created, one in the real estate sector and the other related to financial crime. His knowledge and experience are evident in the way he navigates the reader from the basics of AI uses to the underlying technological aspects, its vast potential being researched and explored, and the cautionary notes about AI's uncontrolled development and use.

The book begins with various examples of how the young are currently being engaged and influenced by AI in their prolific use of social media. He then delves into the various business uses from automation to problem-solving capabilities and finally peeks at a longer horizon of transformation into "superintelligence," where the machines can think beyond human capability. While focusing primarily on the beneficial aspects of AI, the author does not shirk from cautioning the reader at every step on how the uses, interpretations, and intent may lead to negative consequences. It reminds me of another transformative technology from the middle of the 20th century: nuclear power. When used with proper guardrails, controls, and intent, it provides a boundless source of clean energy, while its darker side leads to destructive weaponry. AI can be seen in a similar light. One of the leading AI firms, Anthropic, has been extremely cautious in the release of its flagship product, Claude Mythos, which is a powerful tool for individuals and businesses, but its underlying technology and vast potential can make an explosive mix in the wrong hands. Nishant in his book has skillfully navigated through all the uses and caveats and keeps the reader engaged until the very end. The book has helped me organize my thoughts around AI and I sincerely hope the young student readers, among others, will find it equally beneficial.

**–Dr. Manoj Singh**

Senior Risk Financial Executive, New York, USA.

# TABLE OF CONTENTS

INTRODUCTION

# WELCOME TO THE AI ERA

What if something you use every day quietly understands you better than you think?

From the videos you watch to the answers you search, from the music you love to the content that keeps you scrolling, there's something working behind the scenes, learning from you, adapting to you, and sometimes even influencing you.

That "something" is **artificial intelligence**.

This book is for anyone curious about understanding AI, whether you're in high school, college, or just want to make sense of the technology shaping your world. For students especially, this isn't just another tool you'll use; you're the generation that will shape how AI is built, applied, and understood.

You don't need to be a coder. You don't need to be a tech expert. You just need curiosity and maybe a little skepticism.

## Why This Book Exists

AI is powerful. That's obvious.

But what's not obvious is:

- When AI is helping you, and when it's influencing you
- When it's making your life easier, and when it's making decisions *for* you
- When it's a tool, and when it becomes a risk

Most people are learning AI the same way they learned social media, by using it without fully understanding it.

That's exactly how people lose control.

This book exists to make sure you don't.

## What You'll Walk Away With

By the time you finish this book, you'll be able to:

- Understand what AI actually is (no jargon, no confusion)
- Recognize where AI is helping and where it can harm
- Use AI to learn faster, think better, and stay ahead
- Spot risks like misinformation, bias, and manipulation
- Make smarter decisions about your digital life
- Start thinking about how *you* can build or shape AI

In short, you won't just use AI. You'll understand it.

## Where This Is Coming From

I'm the founder, chairman, and CEO of DigsFact, where we build real-world AI systems solving high-impact problems through products like InstaBud, which uses AI to transform property inspections, and The PreCogs, which uses advanced AI to detect and prevent financial fraud before it happens.

I spend my time thinking about two things: *how AI can be powerful, and how it can go wrong if we're not careful.*

This book is not theory. It's built from what's actually happening in the world.

## Here's the Real Question

AI is not just another technology.

It's a force that can:

- Help a student learn faster than ever before
- Help a doctor detect disease earlier
- Stop fraud before it happens

...but it can also:

- Spread misinformation at scale
- Influence millions of people silently
- Create risks we're only beginning to understand

So the question isn't whether AI is good or bad, but about who understands it well enough to use it responsibly.

Let's begin.

PART 1

# YOUR LIFE, POWERED BY AI

*"Welcome to the real world."*

– The Matrix

CHAPTER 1

# WHAT IS AI, REALLY?

*"Sometimes you gotta run before you can walk."*
– Iron Man

Whether you realize it or not, AI is already part of your everyday life.

Every time you open TikTok, Instagram, Snapchat, YouTube, or Netflix, AI is working behind the scenes. It helps decide which videos appear on your feed, which songs Spotify recommends, and even which ads or products you see online. Many students are also already using tools like Google Gemini and OpenAI ChatGPT for homework, research, brainstorming ideas, and studying.

AI is no longer something from science fiction movies. Rather, it's already built into the apps, devices, and technology we use every single day.

But what exactly is AI, and why is everyone talking about it?

## AI Is Already Everywhere

Today, we encounter AI in various forms, such as self-driving cars like Tesla and Waymo. While AI may seem futuristic, most people are already interacting with it constantly without even noticing.

For example, when you open TikTok, Instagram, or YouTube, AI helps decide which posts and videos appear on your feed based on your interactions. The more you use these apps, the more the AI learns about what you enjoy watching.

AI is embedded in our phones, cars, apps, and websites. Some of the most common examples include Netflix and YouTube, where AI works behind the scenes to recommend movies, TV shows, and videos based on your viewing history. If you have ever seen the "Recommended for You" section on Netflix, that's AI trying to predict which films you may enjoy watching next.

Similarly, Snapchat employs AI for face recognition, filters, and photo effects. AI can recognize facial features, adjust lighting, and apply filters almost instantly.

Another example of AI in everyday life is Spotify. Spotify analyzes the music you listen to and recommends songs or creates playlists based on your taste.

AI also plays a major role in what goes viral online. Platforms like TikTok, Instagram, and YouTube use AI to analyze user habits, what people are watching, and how long they spend viewing certain content. Based on this data, AI decides which videos should be shown to more users.

## So, What Exactly Is AI?

In simple terms, AI is software that learns from data, identifies patterns, and makes decisions or predictions. These decisions can be simple, like recommending a song, or more complex, like helping a car navigate traffic.

If you're curious about how some of these systems work, let's delve deeper into one example: Gemini.

Gemini can engage in conversations, generate text, and even help with research. Many students already use tools like Gemini or ChatGPT to explain concepts, brainstorm ideas, or help write essays and assignments.

It can feel like these tools are intelligent because they communicate in a very human-like way. However, it is important to remember that AI systems learn from massive amounts of data and conversations. They are not actually "thinking" like humans.

## AI Can Be Helpful, But It Can Also Be Wrong

The output you get from Gemini, ChatGPT, or similar AI systems is not guaranteed to be one hundred percent accurate or believable. AI can still make mistakes, provide incorrect information, or sometimes even invent details completely.

This happens because AI is trying to predict the most likely answer based on patterns in its training data. Sometimes those predictions are correct, and sometimes they are not.

As a user of AI tools, it is important not to accept every answer blindly. You should always:

- Proofread responses
- Fact-check important information
- Look for anything suspicious
- Verify information using trusted sources

This becomes particularly important in areas where mistakes can have serious consequences, such as self-driving cars.

## AI in Self-Driving Cars

If a self-driving car made by Tesla or Waymo makes an error, it could result in injuries or accidents. The risks are much higher than simply getting a wrong homework answer from ChatGPT or Gemini.

Self-driving cars operate in unpredictable environments. A pedestrian may suddenly cross the street, weather conditions may change quickly, or construction zones may confuse the system.

These vehicles rely on cameras and sensors to detect objects and predict movement around them. And although this technology has improved significantly, it is still not perfect.

That is one reason why fully autonomous vehicles are not yet everywhere. Tesla's Full Self-Driving system still requires a human driver to supervise the vehicle. Waymo also operates autonomous vehicles in limited cities while continuing to improve the technology through testing and data collection.

## The Different Types of AI

When discussing AI, many people think about movies like *Iron Man, Avengers: Age of Ultron, The Terminator, Minority Report, M3GAN, I, Robot,* or *Mission Impossible—The Final Reckoning.* These movies often show worst-case scenarios where AI becomes dangerous or uncontrollable.

While those stories are fictional, they raise real questions about how advanced AI could affect society in the future.

However, it is important to understand that there are different types of AI.

## Narrow AI (What We Have Today)

The AI we use today falls under "Narrow AI."

Narrow AI is designed to perform specific tasks. Examples include ChatGPT, Gemini, Netflix recommendations, and Spotify playlists.

These systems can be very powerful within their area of training, but they cannot think independently or perform tasks outside of what they were designed to do.

For example:

- Gemini can help with writing essays and answering questions, but it cannot drive a car.
- A self-driving car can navigate roads, but it cannot help you study for an exam.
- Spotify can recommend music, but it cannot help you study for an exam.

Each AI system is trained for a particular purpose.

## AGI: The Next Stage of AI

The next possible stage of AI is called "Artificial General Intelligence (AGI)."

AGI doesn't exist yet, but many experts believe it may emerge in the future.

Unlike Narrow AI, AGI would be able to learn and perform many different tasks, more similarly to how humans think and adapt. An AGI system could potentially help you study, drive a car, build software, hold conversations, and solve problems all within the same system.

Some people imagine future humanoid robots, like the ones discussed by Elon Musk, eventually becoming AGI-powered assistants capable of helping around the house, shopping for groceries, driving cars, and having conversations. However, true AGI has not yet been achieved.

## Superintelligence

Beyond AGI is the idea of "superintelligence," which refers to AI that surpasses human intelligence in nearly every area.

This is the type of AI often shown in science fiction movies. Some people believe superintelligent AI could help solve major global problems such as curing diseases, reducing traffic accidents, or finding solutions to climate challenges.

Others worry that AI that is more intelligent than humans could become difficult to control.

The exciting part about AI is its potential to improve our lives, but there are also significant concerns about safety, misuse, and loss of control. Questions such as:

- What happens if governments misuse it?
- What happens if AI systems begin acting unpredictably?
- Could humans lose control over advanced AI?

These are serious discussions taking place.

## Why Understanding AI Matters

AI is already deeply connected to our everyday lives. It powers platforms like Spotify, Netflix, TikTok, Gemini, and ChatGPT. These tools can help us save time, learn faster, and become more productive, but they can also create risks if used carelessly or irresponsibly.

That is why it is important to approach AI thoughtfully. We should use AI to improve our lives while also understanding its limitations and potential dangers.

**Key Takeaways:**

1. Narrow AI = specialist AI focused on one task (dominates today)
2. AGI / General AI = human-like AI capable of many tasks (possible future technology)
3. Superintelligence = AI beyond human intelligence (theoretical and uncertain)

**Question for you to think about:** What *can* AI do versus what *should* AI do?

CHAPTER 2

# THE COOL SIDE OF AI

*"More human than human."*
– Blade Runner/Blade Runner 2049

In this chapter, I want to dive deeper into the positive aspects of AI. In the previous chapter, we discussed what AI is and briefly touched on both its benefits and potential risks. Now, let's explore some of the exciting ways AI is already making a difference around the world and improving our everyday lives.

AI technology powers many of the applications and devices we use every single day, often without us even realizing it. From the moment we wake up and check our phones to the time we go to sleep, AI is working quietly in the background.

## AI in Everyday Life

Many smartphones and apps, such as Spotify, Instagram, TikTok, YouTube, and Netflix, use AI to improve the user experience. Search engines like Google Gemini, along with tools like Anthropic Claude and OpenAI ChatGPT, are also changing the way people search for information online.

When it comes to online shopping, companies like Amazon and Walmart use AI to provide personalized product recommendations based on your browsing and buying habits.

In fact, we likely interact with AI dozens of times every day without noticing it.

For example:

- Netflix recommends movies and TV shows you may enjoy.
- Spotify suggests songs and playlists based on your listening history.
- TikTok and Instagram decide which videos appear on your feed.
- Amazon recommends products based on what you searched for or purchased before.

AI studies patterns in your behavior and tries to personalize your experience.

## Your Smartphone Is Packed with AI

Our smartphones are basically AI hubs in our pockets.

One example is facial recognition, which allows you to unlock your device using your face. Another major example is your phone's camera. Most people use their phones every day to take photos and videos, but behind the scenes, AI plays a significant role in their enhancement.

AI adjusts lighting, sharpens photos, improves focus, applies filters, and enhances night mode, so photos appear brighter and clearer.

Another example is the keyboard on your phone. Modern devices use AI-powered autocorrect and predictive text features that help complete sentences as you type. AI-powered speech-to-text tools also allow users to speak messages instead of typing them, making communication faster and easier.

In summary, smartphones are packed with AI features that many people use daily without realizing it.

Even if you are in middle school and do not yet own a smartphone, you are still interacting with AI through streaming platforms, games, websites, and social media.

## How AI Learns What You Like

AI constantly studies user behavior to improve recommendations.

For example:

- Did you watch an entire reel or scroll away after five seconds?
- Did you replay a song several times?
- Did you skip a song after the first ten seconds?
- Did you like, comment on, or share a post?

Platforms like TikTok, Instagram, Spotify, and YouTube analyze all of these behaviors to better understand your interests.

That is why your TikTok "For You" page looks completely different than someone else's. AI learns from your behavior and builds a personalized experience specifically for you.

Online shopping works similarly. If many people suddenly start buying a product because it became popular online or went viral, prices may even increase because of higher demand.

You may also notice that after searching for a product online, advertisements for that same product begin appearing on Instagram, Gmail, YouTube, or Facebook. These AI-powered systems are connected and designed to predict what products users are more likely to buy.

## AI Helps Protect You Online

AI is not only used for entertainment and recommendations but also helps protect people online.

For example, spam filters in email services like Gmail use AI to detect suspicious emails and phishing scams. These systems analyze millions of emails and learn to recognize dangerous patterns, helping block harmful or fake emails before users even open them.

Without AI, many more scams and dangerous links would likely reach people's inboxes.

## The Good Side (and Trade-Offs)

On one hand, AI makes our lives faster, easier, and more personalized. But at the same time, it can also create a dependence on technology.

For example, when you log into Netflix and don't feel like searching for a movie, you may simply choose something from the "Recommended for You" or "Top 10" section. Spotify creates playlists automatically based on your listening habits, so many people rely on the algorithm instead of discovering music themselves.

AI can make life more convenient, but it also influences many of the choices we make online.

That is why it's important to understand how these systems work instead of relying on them automatically.

## AI in Gaming and Music

AI is also transforming gaming.

Many modern games use AI to adjust difficulty levels based on a player's skill level. Instead of every player facing the exact same challenge, the game learns from how well each person performs.

For example:

- Beginners may receive easier levels.
- Experienced players may face tougher opponents.
- The difficulty increases as players improve.

This helps games remain running, engaging, and challenging for different types of players.

AI is also changing music. Beyond recommending playlists, AI can now generate music and imitate the styles of different artists. Some AI systems are already capable of creating songs that sound similar to songs made by real artists.

In some ways, AI may even understand your music preferences better than many of your friends do because it constantly analyzes what you listen to, skip, replay, or like.

## AI in Sports

AI is becoming increasingly important in sports like football, basketball, and tennis.

Athletes and coaches use AI to study performances, improve training, reduce injuries, and develop strategies.

For example:

- Analyze player movement
- Identify weaknesses in opposing teams
- Recommend strategy adjustments before games

NFL coaches use AI to study opponents and prepare game plans. Basketball teams also use AI to monitor player health and predict possible injuries before they happen.

The goal is not only to win games but also to help athletes perform at their best while staying healthy.

## AI in Space Exploration

AI is also helping humans explore space.

Organizations like NASA and SpaceX analyze vast amounts of space data using AI systems. AI helps spacecraft and rovers navigate environments, collect information, and make decisions without constant human control.

For example, Mars rovers can move across difficult terrain, collect samples, and send information back to Earth.

AI is also helping scientists discover planets, galaxies, and patterns in space that humans may have missed on their own.

This shows that AI is not merely a buzzword; it is a powerful tool helping humans to solve problems, make discoveries, and push technology forward.

## Why This Matters

AI is transforming entertainment, shopping, gaming, sports, communication, and science. Whether we realize it or not, AI influences many parts of our daily lives.

From the moment we wake up until we go to sleep, AI helps shape what we watch, listen to, buy, search for, and interact with online.

The exciting part about AI is that it has the potential to make our world smarter, faster, more personalized, and more innovative when used responsibly.

**Key Takeaway:**

From your playlist to your game strategy, AI is quietly making your digital world smarter, faster, and more personal.

CHAPTER 3

# AI AND YOU

*"There is no spoon."*
– The Matrix

This chapter focuses on connecting you with AI and understanding how it influences your everyday life.

AI is not just a fascinating technology or something companies talk about online. It's actively shaping your experiences every single day. Whether you are scrolling through TikTok, watching YouTube videos, or listening to Spotify, AI is constantly making decisions behind the scenes.

For example, the videos and posts you see on TikTok or YouTube are not random. AI systems decide what appears first in your feed, what order content is shown in, and what is most likely to keep you watching, scrolling, and engaging longer.

## The "For You" Effect

If you observe people on public transportation, you'll probably notice many people endlessly scrolling through videos and reels. This is not an accident.

AI has become extremely advanced at predicting what people will find interesting. It studies your behavior to determine:

- What type of content you enjoy
- How long you watch certain videos
- What you like or share
- Even what makes you stop scrolling

The goal is to keep you engaged longer, as more time means more revenue for companies.

The more time people spend on apps like TikTok or YouTube, the more money those companies can make through ads, engagement, and data collection.

This is why your feed is personalized specifically for you.

If you regularly watch a certain type of video, you will start seeing more of that content. If you ignore or skip certain topics, the AI will slowly show you fewer of them.

The AI pays attention to almost everything, including likes, comments, shares, watch time, replays, searches, and even pauses while scrolling.

This personalization is often referred to as the "For You" effect. Your feed may look completely different from your friend's feed, even if you are sitting right next to each other.

The AI creates a personalized bubble around you based on your interests and behavior.

## The Bubble Effect

One of the positive sides of AI is that it helps show you content you are likely to enjoy. But there can also be downsides.

Because AI keeps recommending similar types of content, it can slowly trap people inside a "bubble." Instead of being exposed to new ideas, perspectives, or interests, people often continue to see more of what they already agree with or enjoy.

For example, open Netflix and compare your movie recommendations with a friend's. You may notice that their suggestions look vastly different from yours.

There may even be movies or shows on Netflix that you never discover because the algorithm does not think you will be interested in them. Unless someone tells you about those movies or you search for them manually, they may never appear in your recommendations.

This bubble effect isn't just limited to content; it extends to our social interactions as well.

If you mostly interact with reels from a few specific friends, those are the people whose content you will continue seeing most often. Meanwhile, other people you follow may rarely appear in your feed simply because you haven't engaged with their content recently.

This shows that AI influences not just the content you see but also the people you interact with online.

## Your Digital Footprint

Every action you take online contributes to something called your "digital footprint."

This footprint encompasses everything from your searches, likes, and posts to your location data and navigation history when you use Google Maps or other map services.

There are two types of digital footprints.

## Active Digital Footprint

This comes from actions you directly choose to take, such as:

- Liking posts
- Commenting
- Posting photos
- Searching for something online

## Passive Digital Footprint

This happens behind the scenes when apps and devices collect data automatically.

For example:

- Map apps track routes and locations.
- Websites track browsing behavior
- Devices like smart speakers may listen for voice commands.

Companies use this information to personalize ads, recommend products, and predict what users are most likely to engage with.

## Why Companies Collect So Much Data

The main reason companies collect so much information is simple: they want to keep users engaged and increase revenue.

The more AI understands your behavior, the better companies can:

- Recommend products
- Show targeted ads
- Personalize content
- Keep you spending more time on their platforms

In many ways, your attention and time have become extremely valuable.

The more time you spend watching videos, scrolling through social media, or clicking on content, the more profit companies can actually make.

This is why understanding AI matters so much.

## The Positive Side of AI

With any technology, there are pros and cons.

AI can save time, improve productivity, help people make decisions, personalize recommendations, and help users discover helpful content faster.

When used responsibly, it can make life easier and more efficient.

The key is learning how to use AI as a tool without allowing it to fully control your attention, habits, or thinking.

## Privacy, Manipulation, and Addiction

At the same time, there are important risks to understand as well.

One major concern is privacy. Many devices and apps collect huge amounts of information about users. Phones, smart assistants, apps, and websites all gather data that helps build a detailed profile of your habits and interests.

Another concern is manipulation.

Because AI constantly learns what keeps people engaged, it can influence what users watch, believe, and focus on. Over time, this can create an "echo chamber," where people mostly see information that supports what they already believe while rarely seeing different perspectives.

There's also the risk of addiction.

Many people spend hours scrolling through endless reels, videos, and social media posts. Sometimes, people sitting right next to each other are both focused more on their phones than on actual conversations.

This cycle of repetitive content can lead to dependency.

## Awareness Is the First Step

The good news is that we are not helpless.

The first step toward taking control is simply becoming aware of how AI influences us. Once you understand how these systems work, you can begin making more intentional choices about what you watch, click on, share, and spend time doing online.

It's perfectly fine to use AI tools and recommendations. But it is vital to remain conscious of how they may be shaping your decisions and attention.

Another important habit is learning to question what you see online.

As mentioned in the first chapter, AI is not always correct. Just because a video, image, or post appears convincing does not mean it's true.

Today, AI can generate fake videos, fake images, cloned voices, and misleading content that can easily fool people.

That is why "seeing is believing" is no longer always true in the age of AI.

## Think Critically

To avoid misinformation, try to gather information from multiple sources instead of relying on only one platform or creator.

If you see something surprising online:

- Look it up on different websites.
- Check trusted news sources.
- Compare multiple viewpoints.
- Think critically before believing or sharing it.

The goal is not to fear AI, but to understand it.

**Key Takeaway:**

You have a digital footprint, and it's important to be intentional with your online actions. Question what you see, be aware of the potential risks, and make sure to take advantage of the positive aspects of AI, which can enhance your productivity, intelligence, and overall quality of life.

**Key Takeaway for Part 1:**

AI is already part of your everyday life. Understanding how it works is the first step to staying in control of it.

PART 2

# HOW IT WORKS

*"That, detective, is the right question."*
– I, Robot

CHAPTER 4

# WHAT'S INSIDE THE AI BRAIN?

*"I'm sorry, Dave. I'm afraid I can't do that."*
– 2001: A Space Odyssey

Now that we have discussed AI's role in your daily life, you should better understand where it is actively at work. In this section, we will delve deeper into how AI functions.

In other words, what's happening inside the "brain" of AI?

We'll explore how AI trains, whether it truly possesses intelligence, and whether it is as smart as humans using real-life examples that resonate with your everyday experiences.

## What Is Data?

At its core, AI comprises several elements, with data being the foremost. This data is processed by AI models, which are created using "machine learning." If you don't know what machine learning is, it's like teaching a computer by showing it examples instead of giving it step-by-step instructions. The computer looks at lots of examples, finds patterns, and uses those patterns to make predictions. These examples are generally referred to as "data" that is fed to the computer for machine learning.

Data is all around us. It includes everything you encounter on the internet, as well as information about you, including your actions, interests, and preferences.

Data acts as the fuel for AI. It uses various types of data to learn and make predictions, fulfilling its designated tasks.

There are different types of data:

1. Textual Data: This includes your text messages, the books you read, and the websites you visit.

2. Image Data: This encompasses photos you take with your phone or camera and the videos you capture.

3. Audio Data: This consists of the songs you listen to on platforms like Spotify, voice recordings you make, and audio messages you send.

4. Behavioral Data: This includes how you interact with content throughout the day, what posts you click on, the Instagram feeds you view, and the movies you watch on Netflix, whether they are action, comedy, or horror.

In general, for AI to be accurate and effective, it needs exposure to more data. However, simply having a large quantity of data does not guarantee accuracy; it also requires quality.

To illustrate the difference between quantity and quality, consider this example: if you trained an AI only on memes from friends, what would it be like with no other data?

Let's break this down. When we discuss data in the context of machine learning, we typically refer to three essential elements: data, model, and machine learning itself.

## Data

Data comes in two forms: raw data and labeled data.

You can think of "raw data" as a blob of unorganized and unstructured information. It lacks specific tags or labels that describe or categorize its content.

On the other hand, "labeled data" is clearly defined and provides specific descriptions of the data being analyzed. For example, imagine you have a collection of images on your phone. The unstructured data would be all your photos combined. However, if we label all the pictures of cats as "cat" and all the pictures of dogs as "dog," we create labeled data.

A great example of this is the Photos app on your iPhone. The app organizes photos into collections and applies labels based on the people present. So, when you want to find all the pictures featuring your close friend, you can simply tap their name, and all related photos will appear. This labeled data improves your ability to search and find specific images.

Labels are crucial for teaching AI how to interpret various pieces of data. For example, when AI processes numerous images labeled as "cat," it learns to identify what a cat looks like. Without these labels, the AI would struggle to discern a cat from a dog.

Another example involves self-driving car companies, such as Tesla and Waymo. They utilize vast amounts of labeled images, including countless pictures of stop signs, to educate their vehicles on what a stop sign signifies and how the car should respond when encountering one. Without labeled data, the car would lack the necessary understanding and guidance to react appropriately.

## Model

The model acts as the brain of the AI, learning patterns and relationships within the data it's exposed to. It doesn't inherently know the answers or retain them; rather, it identifies patterns in the data.

A helpful analogy is to consider ingredients and recipes. If the data represents the ingredients, then the model serves as the recipe that explains how to use those ingredients effectively. The final output is the finished dish, dessert, or meal.

In summary, data serves as the ingredients, and the model functions as the recipe, and together they combine to create something meaningful, whether it's a dish or an AI system capable of making informed decisions.

But to serve as an effective guide, the model needs to be trained.

## Training a Model

Training is essential for developing an effective model.

"Training" means continuously feeding data into the model. When you input data, the model makes a prediction based on that information. Once a guess is made, the model compares its prediction with the correct answer and adjusts itself to enhance future guesses. This cycle of guessing and refining allows the model to make small tweaks each time.

This process is typically repeated thousands or even millions of times, depending on the complexity of the data involved.

For example, consider the case of identifying stop signs in various contexts. As the model is exposed to numerous images of stop signs, each at different angles, distances, and sizes, it gradually becomes more accurate at recognizing them. Eventually, after many iterations, the model successfully learns to identify them in diverse scenarios.

## Putting It All Together

To summarize, data is essential for training a model, which is a core component of the machine learning process.

There are two main types of machine learning: "supervised learning," when you provide the model with labeled data, and "unsupervised learning," when the model analyzes raw data to discover patterns on its own. In both cases, the model makes predictions, checks them against known answers, and adjusts to improve its accuracy.

The ultimate goal of training is straightforward: to enhance the model's predictive capabilities over time.

However, as with anything, there are tradeoffs.

## The Costs (Literally and Figuratively)

Training is resource-intensive. It requires vast amounts of data, often millions or even tens of millions of data points, as well as significant computational power.

You may have noticed that when you run many apps on your phone, it sometimes freezes or crashes. This happens because the device has limited processing power. When overloaded, it struggles to function properly.

The same is true for training models; if the hardware lacks sufficient capabilities, it cannot handle large datasets effectively.

First, using these computers consumes a lot of energy. This process primarily involves electricity and requires considerable time, including both the time people spend working on it and the processing time needed for computation.

Of course, all of this contributes to the overall expense of training these models, so acquiring training data can be quite costly.

## How It Applies To You

To illustrate how we use different AIs in our daily lives, consider Snapchat or Instagram filters. These applications have been fed millions of images of faces, allowing the AI to learn the positions of various facial features. Because people's facial structures can vary greatly, the AI must learn to adapt the filters accordingly so that they fit well on each individual's face.

If an AI is trained on a limited range of faces, the filters won't work effectively on others. For instance, if the training data mainly consists of North American faces and that same filter is applied to Asian individuals, it may not perform well due to differences in facial features. Therefore, it's crucial for the AI to be exposed to a wide variety of data in addition to a large quantity of it.

There is no magic behind how AI acquires knowledge. It learns from data. The model improves over time as it processes more data, much like a person learns and grows from experience.

Having covered data, models, and machine learning, the next crucial aspect to understand about AI is its ability to recognize patterns and make decisions. "Pattern recognition," as the name suggests, involves AI identifying similarities within large sets of data. It seeks out trends and looks for recurring elements.

This is the essence of pattern recognition.

If there are a million photos, and all of them share a similar feature, like a stop sign shape, then the system will recognize that these images have something in common. It identifies the repeated structures and trends present in the images.

Here are some real-life examples of pattern recognition:

1. Faces In Photos: When you apply a filter in apps like Snapchat or Instagram, the AI compares your photo to millions of others it has been trained on. It searches for similarities in faces to apply the effects accurately.

2. Spam Emails: Email systems are trained on countless spam messages. They look for repeated patterns and structures typical of spam to classify incoming emails, directing suspicious ones to the spam folder.

3. YouTube Recommendations: YouTube predicts what videos you might want to watch next by analyzing your viewing habits. It tracks your repeated actions and preferences to recommend similar content.

4. Auto-Suggest On Phones: When you're typing a message, your phone's auto-suggest feature tries to complete your sentences. It predicts the next words based on what you've previously typed, effectively suggesting the most likely completion without understanding the context.

## What It Actually Does

AI doesn't truly understand the meaning behind your text; it predicts the probabilities of what comes next based on past data. This technique is how chatbots generate responses, and it's also how AI image generators create visuals. They generate outputs by predicting based on what they've previously encountered.

Decision-making in AI relies on pattern recognition, but it also factors in the confidence level of its predictions.

For example, suppose you show the AI a new image of a cat (that isn't labeled as such) after training it on millions of labeled cat images. The AI may not be entirely sure if the new image is a cat, but it will assess the probability. If it determines a ninety percent confidence that the image is indeed a cat, that probability guides its final decision.

There's a process of pattern recognition that informs AI decision-making, which is influenced by various factors. Specifically, the confidence level of an AI's identification, such as determining whether an image is of a cat, depends on the amount, quality, and variety of data it has been trained on.

However, as you may have heard, AI decisions can sometimes be incorrect. Just because AI generates a response doesn't mean it's always accurate.

This inaccuracy can occur for several reasons. For instance, if the data provided is flawed, it can lead to incorrect patterns. Additionally, the model might struggle with situations that don't frequently occur, leading to confusion.

Consider this question: If you had an AI system that was consistently around ninety percent confident in its decisions, would you trust it to handle all decision-making?

**Key Takeaway:**

AI doesn't magically know things—it learns from data, just like you learn from experience. AI decisions are only as good as the patterns it learns, which, in turn, rely heavily on the quality and quantity of the data provided.

CHAPTER 5

# TRAINING AI: WHAT COULD GO WRONG?

*"Everyone creates the thing they dread."*
– Avengers: Age of Ultron

In earlier chapters, we discussed what happens inside the "AI brain" and how AI learns from data and patterns. But this raises an important question:

What happens when the training itself goes wrong?

If an AI system is trained incorrectly or learns from flawed data, the consequences can be significant. AI systems make decisions based on the data they are given, so if the data is biased, incomplete, or unbalanced, the AI itself can become biased as well.

## What Is Bias in AI?

One of the biggest problems in AI is something called "bias."

In simple terms, "bias" means an unfair or unbalanced way of thinking or making decisions. In AI, bias usually happens when the data used to train a system does not properly represent everyone or everything equally.

This can happen when:

- Certain groups are overrepresented.
- Other groups are underrepresented.
- The data reflects existing stereotypes or historical inequalities.

When AI learns from unbalanced data, it also learns the imbalance.

For example, consider filters used on Snapchat or Instagram. If a filter is trained mostly on faces from one demographic, it may not work as well on people from different backgrounds. Features like facial tracking, filters, or lighting adjustments may perform poorly because the AI was not exposed to enough diverse examples during training.

This shows how important representation is in AI systems.

## When AI Learns the Wrong Patterns

A well-known example happened shortly after the launch of OpenAI ChatGPT. Someone asked the AI to describe what a CEO looks like, and the model generated the image of a white male.

Why did this happen?

Because the AI had likely seen more examples of CEOs represented that way in its training data. AI systems learn patterns from the information they are exposed to. If most examples follow one pattern, the AI may incorrectly assume that pattern represents everyone.

This highlights an important point: AI does not automatically understand fairness or diversity. It simply learns from the data it receives.

That is why proper representation across genders, ethnicities, backgrounds, and experience is so important.

## Facial Recognition and Real-World Problems

Another major example is facial recognition technology.

Facial recognition works best when it has been trained on millions of diverse faces from different backgrounds, skin tones, and age groups. If the system is trained mostly on one demographic, it may struggle to recognize people outside that group accurately.

This can lead to serious real-world consequences, like:

- Incorrect identifications
- Unfair treatment
- Systems that work better for some people than others

The issue is not always the technology itself. Often, the problem comes from the data the AI was trained on.

## Hiring Algorithms and Bias

Many companies now use AI-powered hiring systems to help sort through job applications.

These systems can save time for HR departments by scanning resumes and identifying candidates that match certain criteria. However, problems arise when the algorithm is trained on biased historical data.

For example, imagine an AI system trained mostly on resumes from graduates of one university or people from one specific background. The AI may begin favoring applicants who resemble those past examples while unfairly rejecting others.

This means qualified candidates may never even have their resumes reviewed by a real person.

Bias in hiring algorithms can quietly affect thousands or even millions of people without them realizing it.

## AI and Loan Approvals

Another important example is home loans and financial approvals.

Buying a home is one of the biggest financial decisions many people will ever make. Today, AI systems are increasingly used to help review loan applications.

But if the training data mainly includes people from one type of financial background, income level, or neighborhood, the AI may unfairly favor similar applicants while rejecting others.

The dangerous part is that many people assume that machines are automatically objective or fair.

However, AI is only as fair as the data it learns from.

If biased data goes into the system, biased decisions can come out of the system.

## Can AI Repeat Mistakes of the Past?

AI systems often learn from historical data, but this creates a thoughtful question:

If AI learns from history, can it also repeat the mistakes of history?

The answer is yes.

For example, imagine a school using AI to automate disciplinary decisions based only on past disciplinary records from that school. If the historical data already contained unfair patterns or unequal treatment, the AI may continue repeating those same patterns.

Instead of fixing problems, AI could accidentally reinforce them.

This is one reason why human oversight is so important.

## Why Representation Matters

At the center of this issue is one important idea: representation.

"Representation" means making sure different people, backgrounds, experiences, and data types are properly included in the training data.

For AI to work fairly and accurately, it needs diversity in the data it learns from.

For example:

- Some Snapchat filters work better on certain skin tones than others.

- Voice assistants like Alexa, Siri, or Google Assistant originally struggled with many accents because they were trained mostly on certain speech patterns.
- Recommendation systems may ignore niche interests because they focus mostly on what large groups of users watch or enjoy.

If certain groups or data types are missing from the training data, the AI may struggle to understand or serve those users properly.

Better representation usually leads to more accurate and fairer outcomes for everyone.

## AI Does Not Understand Nuance on Its Own

It's important to remember that AI does not naturally understand fairness, context, or nuance the way humans do.

AI cannot automatically recognize who is missing from the data, whether the data is unfair, or whether certain groups are being ignored.

That is why developers must intentionally create diverse datasets and continuously test AI systems across many different groups and situations.

Without careful testing, AI may overlook or misinterpret important information.

## Why Human Oversight Still Matters

This is why many AI systems still rely on what is called "humans in the loop."

This means humans continue monitoring AI systems, reviewing decisions, and stepping in when mistakes happen.

Even very advanced AI systems still require human supervision.

For example, Tesla's Fully Self-Driving technology still requires a driver to sit behind the wheel and pay attention to the road. Cameras inside the car monitor the driver's eyes and behavior to make sure the person is still supervising the system.

AI has incredible potential, but it also reflects the strengths and weaknesses of the data it learns from.

When AI systems are trained carefully using diverse and balanced data, they can become powerful tools that help people. But when training data is incomplete or biased, AI can unintentionally create unfair outcomes at a massive scale.

Before we move on to the next chapter, I pose a simple question:

What happens when your own identity, experiences, or background are not properly represented in the AI tools you use every day?

**Key Takeaway:**

AI doesn't magically remove bias; however, it can quietly amplify it. What AI learns depends on what we teach it, and bad lessons can have real-world consequences.

## CHAPTER 6

# WHY AI STILL NEEDS YOU

*"The past is just a story we tell ourselves."*
– Her

We concluded the previous chapter with a significant point: AI still needs human involvement. Now, let's explore why this is the case.

As we discussed earlier, AI can make mistakes in decision-making due to various factors, and humans play a crucial role in assisting AI to complete and automate many tasks.

A perfect example of the need for humans in the process is the driver's seat of a Tesla, which, despite its fully self-driving capability, still requires a human driver. To understand why AI relies on humans, we must consider what AI is good at and where it continues to struggle.

## What AI Is Good At

AI excels in speed; it can perform tasks much faster than a human can. It also scales effectively from small to large operations, achieving growth rates far beyond human capability. AI benefits from high computing power, which allows for greater output. Furthermore, it is adept at detecting patterns. When fed a large dataset, AI recognizes these patterns and predicts future outcomes based on them.

## Where AI Still Struggles

However, when comparing AI to humans, we must also identify areas where AI struggles.

One key area is context. For instance, consider the autocorrect function when typing a text message. While it sometimes predicts your intended sentence correctly, other times it completely misses the mark, changing the meaning of your message. This is because AI lacks the context and intent behind your words.

Another area where humans excel is ethics. Most people possess a sense of ethics, while AI currently relies solely on pattern recognition from data. As discussed in the first chapter, while the future of AI may involve ethical considerations, current technology struggles to understand and implement ethics.

Humans also outperform AI in unusual situations. A relatable example is using Google Maps. If a road is closed, Google Maps may still direct you along the same path because it hasn't updated its information. I experienced this firsthand while navigating Downtown Chicago on St. Patrick's Day, when many roads were closed, but Google Maps continued to suggest routes blocked by police.

Lastly, the most crucial area where humans have an advantage over AI is common sense. While AI can process information and recognize patterns, it often lacks the intuitive understanding that a human brings to complex situations.

## The Importance of Common Sense

Common sense is the understanding we develop from our experiences. It encompasses the scenarios we frequently encounter in our day-to-day lives. The intelligence we gain from managing these recurring situations is what we call "common sense."

Currently, the types of AI being developed are mostly domain-specific, which means they lack a broader understanding that includes common sense for a variety of situations. Therefore, humans excel above AI at navigating these complexities.

This highlights the importance of human involvement alongside AI, as people can grasp the nuances of specific tasks or situations, question the rationale behind decisions, and challenge them when they seem questionable.

For example, consider a job candidate who appears to be a perfect fit but is rejected by AI because their resume does not match those of previous candidates. A human involved in the process can ask why a seemingly suitable candidate is being dismissed, highlighting the necessity of having humans working alongside AI.

## Real-World Examples of Human Oversight

Here are some scenarios that illustrate the importance of human involvement with AI:

1. Medical diagnosis: AI can suggest diagnoses, but it is the doctor who makes the final decision.
2. Fraud investigation: When an AI flags a transaction as potentially fraudulent, a human needs to investigate further.

3. Self-driving cars: While AI assesses the surroundings, a human must supervise to ensure safety and prevent accidents.

Collaboration between humans and AI is crucial to complete tasks not just effectively, but also responsibly.

## The Risk of Over-Relying on AI

Over-reliance on AI can be problematic. If we begin to trust AI unequivocally, accepting every answer it provides or every decision it makes, mistakes could have significant consequences.

For instance, students may be tempted to use AI to complete their homework. While it's acceptable for them to leverage AI for assistance, they must review the answers for accuracy. If a student blindly trusts the AI and submits incorrect answers, they may face negative repercussions. However, a student who verifies the AI's output will be able to catch mistakes before turning in their work.

A more severe example involves a doctor who relies entirely on AI for prescribing medication. If the AI were to make an error in its recommendation, the consequences could be life-threatening for the patient.

## When AI Makes Financial Decisions

Have you ever used a debit or credit card and had a transaction declined, whether in a store or online? You might have had sufficient funds in your bank account or adequate available credit, but the card was still declined.

If this hasn't happened to you yet, it's good to be aware of the possibility in the future. The reason behind such declines often lies in the decisions made by AI algorithms without human review.

## Ethics, Empathy, and Intuition

Now, let's move on to the next topic: the significance of ethics, empathy, and intuition. It's important to consider why AI needs humans in these areas.

First, let's discuss what ethics means. In simple terms, ethics is about understanding what is right and what is wrong. We learn these concepts as we grow up, through education and family influences. However, AI lacks this moral framework; it strictly follows the instructions given to it. If you instruct AI to perform an unethical action, it won't question you; it will simply execute the command. Therefore, the current iteration of AI has no ethics.

Next is empathy. Empathy involves understanding and sharing the feelings of others. AI does not possess emotions; it cannot experience happiness, sadness, anger, excitement, or any other human feelings. As a result, there is a complete absence of empathy in AI systems. Humans can provide the compassion, sensitivity, and common sense that AI lacks, which is especially important in complex situations.

Another crucial aspect is intuition. Many of us have experienced that gut feeling about certain situations that often turns out to be accurate. This intuition, which can stem from experience or instinct, helps us navigate uncertain or morally challenging scenarios.

## Human Judgment in the Real World

Humans naturally rely on intuition, whereas AI currently lacks any form of intuition.

For example, in healthcare, AI can suggest a treatment based solely on its diagnosis of a disease. However, when a doctor reviews the treatment plan, they consider various factors, including the patient's family situation and the feelings of both the patient and their family members. In this way, a doctor considers much more than just the disease when prescribing a treatment.

Schools and universities are already using AI in education to detect cheating or plagiarism. Unfortunately, AI can sometimes make mistakes and falsely accuse students of cheating when they have done nothing wrong. It's essential to have a human review these accusations to prevent embarrassing situations for innocent students.

In social media and on websites where bots are used, AI can answer basic questions, but it often fails to recognize when a person is feeling frustrated or when a conversation is becoming heated. Unlike AI, humans can quickly adjust their responses based on the emotions involved in the conversation.

These examples highlight the risks of excluding human input or values from decision-making processes. When AI relies solely on data, its decisions can become cold and may even cause harm in certain situations.

This emphasizes the importance of the next generation in shaping the future of AI. As the future builders, decision-makers, and ethical leaders, **you** play a critical role in determining how AI evolves.

**Key Takeaway:**

While AI can enhance our speed and scalability, human oversight is essential for critical tasks, especially when mistakes could be costly or even life-threatening. The future of AI involves not just technology, but also humans. AI will shape the world, but **you** have the opportunity to decide how.

**Key Takeaway for Part 2:**

While AI is undeniably powerful, it is incomplete. It requires better data, improved design, and, most importantly, thoughtful humans behind it. The most effective system combines AI with human wisdom.

PART 3

# BIG POWER, BIG RESPONSIBILITY – THE MACRO VIEW

*"There are no strings on me."*

– Avengers: Age of Ultron

CHAPTER 7

# AI FOR GOOD: CHANGING LIVES AT SCALE

*"Come with me if you want to live."*
– Terminator 2: Judgment Day

Now that we have discussed why AI still needs humans and the importance of human involvement, let's take a broader view of how AI is impacting lives globally. This perspective underscores why humans must continue to play a role in AI development, as it has not yet evolved to the point where we can completely trust it without reservation.

AI has the potential to be used for both good and bad, but its positive applications are already making a significant impact worldwide. For instance, in education, AI is being harnessed to change lives on a large scale.

## AI in Education

AI-based tutoring is one such application. Think of it as your 24/7 homework buddy. Whenever you have a question, you can turn to your AI tutor, which can explain concepts in various relatable ways, making it much easier for you to understand. The best part is that there is no judgment; no one will label your question as "dumb." You can ask the same question repeatedly without worrying about what the tutor thinks of you.

AI tutoring is becoming widespread in many countries, helping students grasp various concepts. Additionally, AI is being used for personalized learning. It evaluates your strengths and weaknesses, tailoring a specific set of questions to help improve your weaknesses while leveraging your strengths.

For instance, if you excel in algebra but struggle with trigonometry or calculus, an AI-based personalized learning system can adjust to your pace and focus on building your skills in these areas. It helps by presenting questions relevant to your current understanding, enabling you to master topics more challenging to you such as calculus.

Beyond math, AI is also helping students write practice essays in both colleges and schools. Many students use AI to generate ideas for improving their essay quality. Furthermore, it is a valuable resource for those learning to code. Students can utilize AI to understand basic coding, create websites, design one-page landing pages, or even develop complete applications with specific features. This allows them to bring their projects to completion much faster than in the past.

## AI and Accessibility

AI is making significant strides in improving accessibility for people with disabilities. For instance, consider visually impaired students. AI enables text-to-speech technology, which converts written text into audio. This means that visually impaired students don't have to miss out on information; they can listen to audiobooks and learn the same material as their peers.

Additionally, real-time translation is helping non-native speakers. In today's global economy, people travel extensively, including students from the United States who study abroad and students from other countries. Previously, language barriers created challenges that hindered students' learning experiences.

However, with real-time translation, teachers and students can communicate effectively, even if they don't share a common language. This technology enables both parties to understand each other and interact seamlessly.

These examples illustrate how AI is making a significant impact worldwide, enhancing our lives and our ability to learn. Students can engage without the stress of feeling inadequate, allowing them to ask questions freely. This makes education more effective.

## AI in Financial Crime Prevention

Now, let's shift our focus to how AI is transforming the financial services sector, particularly in combating financial crime.

AI solutions, such as The PreCogs (https://theprecogs.ai/), monitor your debit and credit card activity in real time. These AI systems act like your own personal "Avengers," working to identify and combat financial crime, such as fraud and money laundering.

They detect unusual behavior in your banking habits and monitor for potential fraud. For example, if there is a sudden large transfer from your account to a location you've never transacted with before, especially at an unusual time, the system flags this activity to help protect your finances.

It's not just a single factor; rather, it's a combination of various elements that help identify discrepancies in behavior patterns. For example, even if someone fakes your ID and appears to pass verification, their transaction behavior may still raise flags. If it doesn't match your usual pattern, the system may flag it as potential fraud, which could lead to the transaction being blocked to protect your finances.

In such cases, financial institutions may reach out to you to confirm whether the transaction was indeed made by you.

You may have received bank texts asking whether a recent transaction was authorized after you used your credit card in a new city or tapped it at a casino for the first time. This is a common security measure: if you're in Las Vegas and use your card, you'll likely receive a text from your bank asking whether the charge was legitimate. You can then approve or decline that transaction.

Essentially, AI is like a digital bodyguard, watching over your transactions.

## AI versus Financial Fraudsters

Fraud detection systems, such as The PreCogs (https://theprecogs.ai/), learn from evolving scam patterns and adapt as financial criminals develop new strategies.

For instance, there has been a rise in romance fraud, a technique that had not been previously encountered. "Romance fraud" is when someone pretends to like or love you online, but their real goal is to steal money, gifts, personal info, or trust.

Think of it as a fake relationship with a hidden scam script. They may send sweet messages and build an emotional connection fast, then suddenly have an "emergency" and ask for money.

Red flags:

- They move too fast.

- They avoid video calls.
- They ask for money, crypto, gift cards, or passwords.

Basically, **if their love story comes with a payment request, it's probably a scam, not a soulmate.**

Fraudsters are continually devising new ways to commit fraud and target vulnerable individuals. Fortunately, modern fraud detection systems can learn and adapt more quickly than these criminals can evolve, allowing them to halt fraudulent activity before it hits the news.

However, it's important to note that AI doesn't operate alone. While it plays a crucial role, fraudsters are also using AI to execute transactions more swiftly and intelligently. They can perform millions of transactions per second, making it impossible for humans to monitor everything on their own.

This is another reason why AI is vital in safeguarding us against financial crimes.

This raises an interesting question about the ongoing race between evolving fraudsters who use AI and AI-driven fraud detection systems. You may have heard the idiom "hindsight is 20/20." It is worth pondering whether AI will ever achieve the level of foresight needed to predict future fraud with the same clarity.

## AI in Healthcare

Now, let's shift our focus to AI's impact on lives globally, particularly in the healthcare industry.

AI is making significant strides in diagnosing diseases and supporting individuals with disabilities and mental health needs. We're already seeing reports of AI detecting tumors in medical scans. There are also instances in which other diseases have been identified at much earlier stages, enabling timely treatment before conditions worsen, often faster than traditional doctors can detect them.

Significant research is currently underway to improve the diagnosis of more severe diseases, such as cancer.

Mental health support through AI chat tools offers coping strategies to individuals facing various challenges. These tools are available 24/7, providing a consistent companion unlike traditional human support services, which operate on specific schedules.

While AI can be an invaluable resource, it should not replace real therapists. Human professionals are essential in the healthcare industry, and AI should be viewed as a supplementary tool rather than a replacement for medical experts.

## AI for Disabilities and Communication

Also, consider the benefits of AI in accessibility. There are AI-based speech-to-text tools that assist hearing-impaired students. These students can read written content during live instruction or from audiobooks, ensuring they have access to the same information as their peers.

Conversely, there are also tools for the visually impaired that convert text to speech, allowing them to listen to audiobooks.

Additionally, AI can describe images for individuals who have difficulty seeing colors or details, enabling them to better understand visual content.

Another application of AI is its potential to help individuals who are unable to communicate verbally.

For instance, Neuralink is a company that is researching ways to interpret thoughts and facilitate communication for those who cannot speak or move due to physical impairments. They are working on innovative technologies to empower these individuals to express themselves.

## AI in Insurance

Now, let's shift our focus to another industry: insurance. Whether it's car insurance or home insurance, understanding these necessities is crucial. If you've recently started driving, you probably know the importance of having car insurance.

The insurance industry is experiencing a significant impact from AI-based tools like Instabud (https://instabud.ai/). These tools enable faster claims processing when incidents occur, such as car accidents or hailstorms that damage homes.

As a car owner or homeowner, filing a claim means you won't have to pay for repair costs yourself.

AI tools like InstaBud can analyze photos and videos of the damage and generate inspection reports within minutes. InstaBud detects damage and materials and includes the damage assessment, potential root cause, severity, repair recommendations, and more in the report it generates.

This technology allows insurance companies to quickly define the scope of work, provide estimates, and expedite repairs, helping you return to your normal life as soon as possible. With tools like InstaBud, the entire claims-processing cycle is accelerated to support individuals recovering from these disruptions.

Another application of AI in the insurance sector is streamlining the process of shopping for new home or car insurance. If you or your parents have ever considered switching insurance providers, you (or your parents) may have found the process time-consuming, requiring hours spent comparing quotes across multiple websites and negotiating with brokers.

AI simplifies this by quickly analyzing images and generating quotes, effectively matching you with insurance carriers faster than ever before.

Additionally, AI tools can be used to compare property conditions before and after rental periods, particularly for short-term rentals like Airbnbs. For instance, AI can assess a property's condition after a stay. This can help identify any new damage that may have occurred.

## AI and Fraud Detection in Insurance

AI tools, such as The PreCogs, also aid in fraud detection within insurance claims.

Fraudsters often attempt to submit fake damage photos, creating realistic-looking images of damage to their vehicles or homes to deceive insurance companies into paying out claims.

Such fraudulent attempts are already underway, and some scammers even intercept insurance payouts, diverting funds from legitimate claimants to their own accounts.

Fortunately, AI tools like The PreCogs are designed to detect these fraudulent activities and prevent them.

## AI and Wildfire Prevention

Beyond the insurance industry, AI is also making a significant impact in sectors like wildfire prevention.

You may have heard about wildfires in places like California or Hawaii; AI is being employed to help address these critical issues.

AI is being used to detect fires early using satellite imagery, sensor data, and weather data. This technology predicts where fires might ignite and identifies areas at a higher risk of rapid spread and fire-related damage, not just to property, but also to human lives.

AI helps answer critical questions, such as where a fire could spread so quickly that people might not have time to escape once it starts.

By predicting these scenarios worldwide, AI can save lives by providing early alerts and ensuring faster evacuations. The goal is simple: to give people enough time to evacuate before a fire reaches them.

What if AI could even detect fire-related dangers long before humans notice smoke? That's where we're heading.

## AI and Energy Efficiency

Another industry example I want to share relates to energy efficiency.

There are growing concerns about climate change, and we need smarter and more efficient ways to use energy while minimizing waste and conserving resources.

In many homes, you might have already encountered smart gadgets that manage lighting based on motion sensors or your presence. For example, devices like Alexa and Google Nest can control your home's lighting.

Smart thermostats have also become more intuitive; they can be based on whether someone is home or away. If the thermostat senses that no one is home, it will lower the temperature to avoid wasting energy heating an empty space. The same principles apply in the summer for cooling.

Smart homes use automated heating and cooling adjustments informed by factors such as occupancy, sleep patterns, and external weather conditions, including whether it is cloudy or sunny.

Additionally, smart cities are investing in AI to alleviate traffic congestion. These systems can predict which areas are prone to heavy traffic and take proactive measures to reduce it. Similarly, AI is optimizing energy usage in both residential and commercial areas.

## AI in Data Centers and Sustainability

Major companies are also leveraging AI to minimize waste in their data centers. As AI technology continues to expand, so does the size and energy consumption of data centers.

A great example of this is how Google, back in 2016, managed to reduce its data center cooling energy usage by up to forty percent. Given the vast amount of data these facilities process, they generate significant heat, necessitating continuous cooling.

By utilizing AI, companies can reduce their energy consumption and contribute to a more sustainable future.

**Key Takeaway:**

When these small improvements are applied at scale, involving millions, or even hundreds of millions, of people, the overall impact can be enormous for everyone. AI isn't just for cool apps and viral videos; it can actually change the world for the better.

CHAPTER 8

# THE DARK SIDE OF AI

*"Everybody runs."*
– Minority Report

In the previous chapter, we discussed the positive aspects of AI and how it is impacting lives globally. However, like everything, AI has both positive and negative aspects. In this chapter, we will examine the negative aspects of AI in more detail.

We have touched on some dangers of AI, but now we will delve deeper into them. You may have heard about deepfakes, misinformation, and manipulation being carried out using AI.

## Deepfakes: When Fake Looks Real

So, what are deepfakes, exactly?

The term "deepfake" refers to AI-generated videos, voices, audio, or images that appear real but are, in fact, fabricated. These media do not depict real events or statements; they are entirely artificial creations designed to look believable.

This can be extremely dangerous, as deepfakes can be used to attribute false statements to individuals or to show them engaging in actions they never actually performed.

You might have seen news stories about fake celebrity videos that portray stars doing things they never did. Similarly, fake political speeches can be created that misrepresent what politicians actually said, tailored to serve various agendas.

On a personal level, consider the dangers of someone using AI to impersonate a parent or friend over the phone.

Imagine receiving a call from someone pretending to be your parent or a friend, inviting you to meet at a specific location. If you go, you might be meeting someone entirely different with ulterior motives.

There are also fake school announcements and prank videos circulating, in which individuals use AI to generate false content that could harm others. The use of AI to create fake evidence in court is another alarming example of its dark applications.

These instances illustrate how deepfakes, essentially fake videos, images, and audio, can negatively impact individuals and society.

## Misinformation and Manipulation

Now, let's discuss misinformation.

In the context of AI, misinformation involves leveraging deepfakes, fake news, or fabricated images to spread false information through various media, including social media.

Countless fake posts on platforms like Instagram and TikTok can proliferate quickly, misleading people and contributing to the spread of misinformation.

Thousands, if not millions, of videos and images can be sent instantly using AI. If these are not regularly checked as part of a verification process, it can become nearly impossible to stop their spread.

This poses a serious danger because we, as humans, tend to trust the videos and photos we see. We believe they are real. However, AI has the capability to easily undermine that trust because it can generate content that looks very authentic.

The critical question then becomes: if you can't trust what you see, what can you trust?

## How AI Shapes Your Beliefs

We previously covered the topic of social media manipulation. Essentially, AI analyzes what evokes emotions in you, understanding your patterns and actions to determine what keeps you engaged or excited. It then floods your feed with similar content.

For instance, if you watch two or three videos on a particular topic, the algorithm recognizes your interest and subsequently fills your feed with more content on that topic. This can lead you to believe that everyone shares your perspective because you are seeing so much of that specific viewpoint.

AI reinforces this perception as your reality.

This phenomenon has a profound psychological effect on individuals, creating a confirmation bias. You tend to believe whatever aligns with your existing views because AI continually presents content that supports those beliefs.

Consequently, you may come to think that everyone shares your views, as AI repetitively showcases that same perspective. This creates a vicious cycle in which AI quietly convinces you to maintain or adopt beliefs you might have begun to embrace.

## Surveillance and Loss of Privacy

Additionally, there's a darker side to AI related to surveillance, social scoring, and an overall loss of privacy.

For example, consider surveillance technologies such as cameras that feed data to AI systems to track your every movement. This includes satellite images and CCTV cameras equipped with facial recognition technology.

Many cameras, whether in public areas or in private companies such as casinos, banks, and federal buildings, or even at city intersections for traffic monitoring, feed data into AI systems.

Facial recognition technology tracks individuals' movements, location, time, and other details.

This means there is a significant risk of surveillance occurring without your consent.

## Social Scoring and Tracking

Social scoring can arise from your behavior across various apps on your phone, whether you're using music, movie, or social media platforms like

Instagram or YouTube. Every action you take, what you listen to, what you watch, how much time you spend on each app, and where you engage by liking or commenting, is monitored.

This data isn't just used to enhance your feed; it's also employed to evaluate how risky you might be for things like loan approvals.

For instance, when you apply for a loan to purchase a car or home, lenders can track your online behavior to assess whether you are likely to repay the loan. They consider factors beyond your credit score, including your social security number and financial history.

Each day, various apps on your phone may already be tracking your location, even if you didn't fully understand the permissions you granted when you installed them. You might be surprised by how many apps consistently monitor your location and activities.

For example, when you search for something on Google, it's not only Google that knows your search history; other apps can also access that information, depending on the permissions you accepted. This is why you might notice ads on Instagram related to your recent Google searches.

## Smart Devices and Privacy Concerns

If you have smart devices in your home, such as Alexa or Google Nest, or use Siri, these devices are always listening to your conversations. They pay attention not only when you actively use their wake words but also during your everyday discussions.

While targeted ads aren't the problem, the potential for misuse by bad actors exists if they can listen to all your conversations. By "bad actors," I am referring to people or groups who use technology in harmful, dishonest, or illegal ways. They might try to steal your data, spy on your conversations, try to scam you, spread misinformation, hack devices, or misuse private information for money, control, or manipulation. This raises significant privacy concerns about the extent of surveillance happening in our daily lives.

There's always a trade-off between the convenience of technology and privacy. Therefore, it's essential to understand what information we share through technology.

## AI and Job Loss

Another critical area where AI has garnered significant attention and concern is job loss.

The question of whether machines will replace humans is a hotly debated topic. Many people have differing opinions on how AI will develop.

Automation of repetitive jobs had already occurred well before current AI advancements. The first tasks typically automated are simple, repetitive jobs that do not require complex decision-making.

Roles such as data entry, customer support, and basic content creation have already been automated using AI. Often, when you think you're speaking to a human in customer support, you're actually interacting with AI.

For jobs like data entry, a task that once took hours or days, AI can process the same information in minutes.

In terms of content creation, many students use AI to enhance their essays and generate new ideas. AI is also capable of creating art. While it may not be genuinely creative, it can generate artwork based on existing knowledge.

## Adapting Instead of Competing

When considering the impact of AI on jobs, it's important to view the situation from a positive perspective. It's not all negative.

Historically, for example, during the Industrial Revolution, while some jobs were phased out by machines, new jobs were created, and existing roles evolved. Therefore, it's not merely a matter of AI taking our jobs; it's about the transformation and evolution of work.

The central concern is whether someone who *utilizes* AI could replace you. For this reason, you must ask yourself how you can learn to leverage AI effectively to enhance your work.

New jobs are created by using technology to perform the same job rather than relying solely on traditional methods.

**Key Takeaway:**

While AI is indeed powerful, it can cause significant damage if we do not control it. We need to be aware of AI's impact. It's essential to establish checks and balances on everything AI does around us, at least for now, until we reach a more responsible phase of AI development. Power without control can result in serious consequences.

CHAPTER 9

# WINNERS AND LOSERS IN THE AI AGE

*"Choice is an illusion."*

– The Matrix Reloaded

We discussed AI at scale, touching on both its positive and negative aspects. Now, let's consider who the winners and losers will be in the age of AI.

It's not just about the desire to learn how to use AI; it's also about your ability to do so.

## Unequal Access to AI

Access to AI tools is not equal across the globe. Many individuals do not currently have the same opportunities to utilize AI.

Within the same city or neighborhood, high-end tools may only be accessible to those who can afford them, while others have to rely on free tools. Additionally, internet limitations play a role; in many rural areas or places with poor cellular signals, access to fast internet can be scarce, and this affects how people can use AI tools.

This inequality can create a discrepancy in opportunities; those with access to AI can gain a competitive edge, while those without it face significant challenges.

For instance, students who can effectively use AI tools may learn and work more efficiently, but this relies on the condition that they use these tools appropriately.

In contrast, students without access to AI are at a disadvantage. They must exert additional effort to keep up with their peers who have access to these tools.

This issue of inequality extends beyond individual schools; when viewed from a global perspective, some countries have widespread access to AI, allowing their populations to find more efficient ways of working.

I have also mentioned various industries where AI is having a positive impact, such as healthcare, insurance, and financial services, particularly in countries with robust access to AI.

On the other hand, there are many countries that lack development, do not have appropriate regulations, or have governments that restrict AI use for various reasons. These limitations hinder their ability to compete on a global level.

Access to AI makes a significant difference, but ultimately, it's important to also consider how effectively it's used.

## AI in Schools: Fair or Unfair?

I encourage you to think about whether AI tools should be permitted in schools. Do you believe they create unfair advantages for some students with access compared to those without?

One school may decide to allow AI in its classrooms, while another may choose not to.

This raises questions about fairness:

- Is it fair to students?
- Does it give an unfair advantage to students in schools that do allow AI?

These are important questions society is still trying to answer.

## Who Is Making the Rules?

Now that we're discussing the implications of AI, one key question to consider is who is truly making the rules. And if you aren't part of that process, are you at risk of being left behind?

We often hear about companies like OpenAI, Anthropic, and Google. These major tech companies are developing AI tools that are being adopted across various industries.

These companies have significant influence over how AI systems are designed and implemented. They embed the rules within their products, which gives them control over the usage of these AI tools.

## The Role of Governments and Regulation

Another influential player in this landscape is the government.

Its role is to regulate the use of AI and manage the associated risks. However, the challenge is that governments often move more slowly than technology evolves. As a result, the negative consequences can occur before regulations are implemented.

We've seen numerous stories of fraudsters using AI to commit crimes, such as stealing money or laundering funds. In many cases, technology is advancing at a pace much faster than the government's regulatory responses, allowing bad actors to exploit these innovations more quickly than regulations can address them.

But once the regulations and guardrails are in place, they pave the path for companies to adopt AI to do constructive development and shape the future that protects against the bad actors, like fraudsters, who use AI for destructive purposes. These regulations also make it harder for bad actors like fraudsters to use AI for hurting others. So, the government plays a crucial role by establishing rules regarding AI usage and defining what AI can and cannot do.

## Researchers and the Future of AI

Researchers also hold significant influence in this area.

Consider the research and development taking place in fields like healthcare, where AI is being used to predict early signs of cancer, or in space exploration, where AI helps utilize resources from other celestial bodies as Earth's resources become depleted.

These examples highlight the extensive R&D efforts across multiple industries, including finance, insurance, healthcare, and more.

These researchers, focused on solving today's problems and envisioning future solutions, can shape the rules to create a better outcome for society.

Those who make the rules have the authority to decide what AI can and cannot do.

Often, regulation lags behind the rapid evolution of AI technology. This discrepancy raises concerns about the potential pitfalls and whether rule makers are keeping pace to ensure AI is safe.

We've seen this play out on social media, where damage has occurred before regulations were established.

## The Risk of Exclusion and Bias

There is always a risk involved in having a limited group of people determine the capabilities of AI. Those not in the decision-making process have little influence over these crucial decisions but are still affected by the outcomes.

A common example is the issue of bias in AI. If AI is trained predominantly on data from a specific region or country, the decisions made by those controlling the data can lead to a system that only reflects that culture, failing to account for others and resulting in biased responses.

This highlights the ongoing risk of exclusion; those who are not involved in decision-making have no say in how AI affects them.

Furthermore, decisions made by a few individuals can influence billions of people, essentially everyone in the world using AI today. This raises a critical question of whether companies, governments, or the general public should determine the rules governing AI?

It's vital to consider who is shaping the future of AI, as their choices will impact the world we all will live in tomorrow.

## The Global AI Arms Race

Now that we've discussed who makes these rules, another significant topic is the global conflicts and the AI arms race that is currently underway.

Countries are competing to develop superior AI technology, as advancements equate to power. With better technology, nations can not only improve their defense capabilities but also enhance their arsenal.

Data is a key factor in this competition; nations are vying for access to data while attempting to restrict others from accessing theirs. This competition is crucial because data, whether raw or labeled, serves as a fundamental building block for AI development.

Consequently, having access to more data provides a significant advantage.

AI represents power because it is transforming various industries and, in turn, impacting our economy.

## AI, Security, and Cyber Risks

We've discussed financial crime, but there are also cybersecurity risks, such as hacking of secure systems and federal records, being carried out by other countries using AI. Moreover, AI is driving innovation; many nations are finding new ways to accomplish tasks more quickly, cheaply, and effectively.

As a result, AI is becoming synonymous with power, which explains why countries are competing in this technology.

There are numerous applications for AI in defense. We continually hear news about one country targeting another's systems to steal data or infiltrate secure records. Drones are being utilized in defense, leveraging AI technologies. The entire intelligence community is using AI to create stronger defenses for their countries, financial systems, and overall infrastructure.

## The Danger of Faster Decisions

However, the real risk associated with applying AI in these areas is that it can expedite decision-making processes.

If there isn't enough human input in the decision-making loop, and if proper controls are not established, AI could escalate conflicts between countries rapidly. While quicker decisions can be beneficial, it's equally important to maintain control over those decisions to avoid unnecessary escalations that could lead to wars and destruction.

It's comparable to playing video games. Many games provide options that allow you to act quickly. If you make a wrong choice, you can hit a reset button and start over.

But in real life, there is no reset button.

**Key Takeaway:**

AI is reshaping power, opportunity, and fairness. AI won't impact everyone equally, and understanding it may become one of the biggest advantages you can have.

**Key Takeaway for Part 3:**

AI has the potential to save lives and improve our world, but it also poses serious risks. Ultimately, AI itself is neither inherently good nor bad; It becomes what we, as humans, choose to make it. You will be the ones steering the future of AI and guiding it in the right direction.

# PART 4
# WHAT YOU CAN DO

*"The Entity's future, or no future at all."*
– Mission: Impossible – The Final Reckoning

## CHAPTER 10

# THE DO'S AND DON'TS OF AI USE

*"Free your mind."*
– The Matrix

In the last few chapters, we discussed the global impact of AI, including its positive contributions, the exciting advancements it's making, and its influence on various industries. We also touched on the darker side of AI, exploring how it can be misused to harm others. Additionally, we examined the concept of winners and losers in the AI age, highlighting issues of inequality, such as who gets to make the rules and who is represented in the conversation.

In this chapter and the upcoming ones, we will focus on what you can do to shape the future of AI. You, as the next generation and future decision-makers, hold the responsibility to influence how AI develops.

The first topic we'll cover is the dos and don'ts of using AI. This is where you will realize you are not just watching the AI story unfold, you are writing it.

So, let's dive in.

## Using AI to Learn versus Using AI to Cheat

When using AI, the most fundamental aspect is your approach: you can choose to use it for learning or cheating. The choice is yours.

Some students already utilize AI as a learning tool, while others misuse it to gain unfair advantages. But what does it mean to use AI for learning?

Instead of seeing AI as a shortcut to complete tasks for you, consider it both a coach and a highly intelligent assistant. This perspective allows it to help you understand complex concepts and break down difficult problems into simpler, more manageable parts. AI can also use language that resonates with you to clarify challenges, enabling you to practice, iterate, and strengthen your skills over time. By using AI as a tutor or coach, you can enhance your learning experience effectively.

## Smart Ways to Use AI

Some examples of effective uses of AI for learning include asking it to explain a math problem step by step or to help simplify a concept for better understanding. Instead of simply requesting the solution, you are expressing that you don't comprehend the problem and are seeking a breakdown to make it easier to grasp.

Other useful applications of AI involve generating practice questions, which can be beneficial for preparation. For instance, if you have written a draft of an essay, you can request feedback from AI to identify any gaps that you might want to address.

AI can also assist in brainstorming ideas. If your teacher assigns a project and you and your friends are unsure about which topic to choose, you can use AI to help generate topics or ideas for your essay.

In all these examples, it's crucial to remember that the thinking process remains yours. You aren't simply handing over the problem to AI and expecting it to solve everything for you. Instead, you are using AI as a tool to assist you in understanding problems, answering questions, and receiving feedback on your writing.

## What Cheating with AI Looks Like

On the other hand, using AI for cheating looks entirely different. In this case, there is no thinking involved on your part. You provide the problem and copy the answers directly. For example, you might ask AI to write an essay for you and then submit it without any modifications.

Similarly, if you are in an exam and use AI for answers, whether or not this is allowed, you are not engaging in the learning process.

Ultimately, the choice is yours: you can use AI to enhance your learning and skills, or to cheat. However, if you opt to cheat, the biggest loser in that scenario will be you.

In the long run, this choice will have the most significant impact on your personal growth and education. Although you might not recognize it immediately, bypassing the learning process deprives you of opportunities to develop essential skills. As a result, you could fall behind your peers who are using AI as a learning aid rather than a shortcut.

The real question you need to ask yourself is how you want to use AI. While it can certainly do your homework, relying on it in that manner may hinder your learning and progress. Think carefully about your approach.

## Long-Term Consequences

You might get away with missing one exam or two, but eventually, an abundance of exams will come, not just in schools and colleges but in real life as well.

Consider students who use AI to cheat. Over time, this behavior can become a habit. They start relying on AI for every assignment, which means they don't truly understand the material. They are simply copying and pasting without engaging in any critical thinking.

When it comes time for exams, they struggle. But it doesn't stop there. After graduating, they face job interviews, where they encounter similar difficulties. As a result, they miss out on ideal job opportunities. In contrast, a fellow student of the same age, who has utilized AI as a learning tool rather than a cheating aid, excels in interviews and secures well-paying jobs.

This presents just one possible scenario.

## A Classroom Debate Worth Having

A suggestion for readers is to propose a debate in your class about the use of AI for homework.

Is AI intended for doing the homework or assignment for you, or is it meant to facilitate smart learning, enabling students to produce better work?

This could lead to a fascinating discussion that reveals various perspectives. You'll discover differing opinions on whether AI is more often considered a tool for cheating or a resource for genuine learning. However, it's important to use AI to *become* smarter, not just appear smarter.

## Using AI-Generated Images, Audio, and Video Responsibly

Now, let's talk about using image, voice, and video generators responsibly.

Generating images, audio, or video with AI is not inherently wrong. However, as highlighted in previous chapters, there are consequences to be aware of.

So, how can you use these tools responsibly?

These technologies can easily produce content that mimics celebrities, friends, teachers, and family members. You might enjoy experimenting with AI to create artwork or visuals for a school project or to generate interesting or funny memes to share with friends. Harness your creativity to explore these possibilities.

However, be cautious about the implications of how AI-generated content, whether it's an image, video, or audio, is used.

## Consent Matters

For instance, generating someone's voice without their permission and using it to create audio content that others hear can lead to sharing misleading or damaging information.

Consider this example. If you create a fake voice of a friend saying something embarrassing, it may seem like harmless fun to you. But ask yourself if you'd be comfortable if someone did the same to you. Cloning your voice to say something embarrassing and distributing it publicly is a violation of consent.

There is a risk to your digital reputation. Anything you create and post online can spread rapidly. It's essential to think critically about how AI-generated content is used and its potential impact on others.

## Once It's Online, It Spreads Fast

When you post something on your feed, at work, in your reel, or anywhere online, it can spread quickly. Additionally, once something is on the internet, it tends to remain there almost indefinitely.

Retrieving or removing a post is challenging; even if you delete it, it may still exist somewhere online. Someone might have liked it, saved it, or shared it, and you have no control over those actions.

For these reasons, it's not easy to undo what you've posted online, which puts people's reputations at risk every time you decide to create and post content without their consent.

## Legal and Ethical Boundaries

This is just one of the less serious aspects of generating content using AI improperly. There are also legal and ethical boundaries that you must not cross, as they can have severe implications.

For instance, creating content without the necessary consent, whether from students, teachers, or others, could violate school or college rules or even break the law.

It's essential to be cautious when producing or sharing content without the consent of those involved.

**Key Takeaway:**

Use AI to learn, experiment, and explore your creativity, but do so responsibly, as your digital actions have real-world impacts.

CHAPTER 11

# BE AN AI SUPERHERO (NOT A VILLAIN)

*"With great power comes great responsibility."*
– Spider-Man

Just like any technology, AI can be used for both good and harmful purposes. The key consideration is how to embrace the role of an AI superhero rather than a villain.

To illustrate this, let's revisit an example we touched on in an earlier chapter regarding the impact of AI in the financial services industry.

## Fighting Fraud with AI

We mentioned how fraudsters are causing harm to ordinary people, affecting not just us, but also our parents and grandparents.

These criminals target vulnerable individuals, stealing their hard-earned money without their knowledge. They leverage AI to stay one step ahead of everyone else.

Enter solutions like The PreCogs. These systems are designed to combat issues like fraud and money laundering.

As fraudsters constantly evolve, they devise new methods to scam people, whether by impersonating others, threatening individuals, or employing other deceptive tactics.

AI-based solutions like The PreCogs help detect and prevent fraud by identifying unusual patterns, suspicious behaviors, and any manipulative actions aimed at customers. They analyze whether the intent of a transaction aligns with the user's typical behavior, employing sophisticated techniques to thwart fraud before it occurs.

What makes these solutions particularly powerful is their ability to learn from every attempted fraud.

The PreCogs can even predict new crime patterns that have not emerged yet but may emerge in the future, allowing us to be better prepared to stop fraudulent activities as they arise.

For instance, when you use a credit or debit card and a fraudster tries to use it in an unfamiliar location, The PreCogs can instantly flag that transaction, preventing the money from leaving your account.

## AI Superheroes versus AI Villains

The same technology is utilized by different entities with different intentions.

While fraudsters act like villains, aiming to harm and steal from individuals, AI solutions like The PreCogs serve as your AI-based protectors, fighting against financial crime such as fraud and money laundering.

When it comes to AI and financial crime, the best strategy is not merely to react to situations as they arise but to predict potential threats and be ready to counteract them. When implemented effectively, solutions like The PreCogs can help protect individuals, save billions of dollars, and foster trust between consumers and their banks or insurance providers. Ultimately, this trust extends to all those who offer essential services to the public.

## Your Turn to Build Something

Now that we've discussed AI's positive and negative uses, let's focus on how you can prepare to be an AI superhero.

It's time to think about a small project, a solution to a real-world problem that does not exist yet, that can help your school or community.

Consider what your very first project could be that goes beyond your regular schoolwork or college assignments, while still addressing a real problem within your school or community.

If you think you can't do an AI project because you don't know how to code or aren't technical enough, those are just excuses. First of all, you don't need to be an expert. Secondly, there are many tools available today that are more accessible than ever.

## Simple AI Project Ideas

If you decide to pursue a simple project, begin brainstorming ideas.

For example, you could create:

- An AI-based study assistant for your classmates

- A chatbot that answers frequently asked questions from new students at your school or college
- A mental health check-in tool to help students who may be struggling
- A homework help platform that promotes smart learning rather than enabling cheating

These examples are just meant to spark your creativity.

If you need more inspiration, think about any problems you've encountered in your school or college during discussions with friends or family. Reflect on moments when someone might have said:

- "I wish our school did this differently."
- "I wish we had this tool."
- "I wish this problem were easier to solve."

Each time you express a desire for change, you have a potential project idea.

If you haven't thought of any issues yet, start paying closer attention. The next time you hear someone in school mention a desire for something different, that's a great opportunity for you to identify a project idea.

Recognizing these situations will help you pinpoint your project focus.

## Community-Based AI Ideas

If you want to engage with the community in a focused way, here are some simple project ideas you may consider:

1. Local Business Recommendations: Explore the existing tools for recommending businesses, but challenge yourself to create a better version.

2. Traffic and Parking Solutions: Traffic and parking issues are common in cities, towns, and around colleges and schools. This could be a valuable area to address for your community.

3. Environmental Tracking: If pollution is a concern in your area, or if residents are actively discussing environmental issues, developing a tool for environmental tracking could benefit the community.

If these ideas do not resonate with you, think about the complaints you frequently hear from your family, neighbors, or other community members.

The next time someone expresses frustration about a specific issue, you may have just identified a potential project focused on community improvement.

## How to Start Your First AI Project

Once you identify a problem to tackle, follow these high-level steps to get started:

1. Identify the Problem: Begin by clearly defining the problem you want to address.

2. Explore Solutions: Consider multiple ways to solve the problem. There is rarely just one right answer; many problems can be approached in several ways. Brainstorm to find solutions that are quick, effective, and easily adaptable.

3. Leverage AI: Once you have narrowed down potential solutions, think about how AI can assist you. Work backward from your desired end state to determine what you need to reach that goal. This will help you outline the steps necessary to move from start to finish.

4. Build a Simple Version: Create a basic version of your solution. It's often much simpler than the final product will be, but it should effectively address the problem without unnecessary features. In the startup world, this is referred to as a "minimum viable product," or MVP. The aim is to design something that can be developed quickly and tested efficiently.

5. Test Your Solution: Once your basic version is ready, test it with those who frequently face the problem you're addressing. The primary goal during this testing phase is to learn from the users' experiences.

6. Iterate and Improve: Use the feedback from your tests to refine and enhance your solution. This iterative process is crucial to developing a successful product.

By following these steps, you can create impactful solutions that address real community needs.

## Start Small and Improve Over Time

Every time you make improvements, it's essential to test your work again. This iterative process allows you to move closer to your desired end state, but remember that achieving that end state will never happen all at once.

Start by building something quickly and simply. Get it out there and test it.

Iterate. This key mindset is crucial for your project. Begin small.

Avoid embarking on a project that will consume a year of your time and energy, especially if it may not be the right solution. If you're going to fail, aim to fail fast.

This means that the initial solution might not be ideal, or it may require adjustments. You'll learn from your first or second tests, which will help you make necessary changes and modifications.

## Focus on Solving Real Problems

The second aspect of this mindset is identifying and solving real problems.

Instead of working on projects that lack impact, focus on genuine issues around you. You don't need millions of users for these projects.

In fact, worrying about having a vast user base should be the last thing on your mind. Instead, pinpoint one real problem and determine how you can address it.

There are numerous teen-led AI projects happening globally. Students from various countries are building AI applications to tackle important issues, such as detecting fake news, creating tools to assist visually impaired individuals, and even developing climate tracking systems.

These examples demonstrate that age is not a barrier when it comes to addressing real-life problems.

You can start with a simple project aimed at solving one of these issues. Remember, impact is more important than complexity.

You don't have to choose the most complicated problem. Instead, select the problem that has the potential for significant impact.

Reflect on this: all those students creating innovative tools and applications around the world are just like you. They made a choice to learn about and utilize AI, positioning themselves as AI superheroes.

What problem in your life bothers you enough that you want to solve it?

You might want to think about this throughout your day, from the moment you wake up until you go to bed. Every time something frustrates you, you may have found an opportunity.

**Key Takeaway:**

AI is not merely a tool you use for getting help with your assignments or inside apps like Netflix or Spotify. Should you choose, it is something you can use to create and build yourself.

# CHAPTER 12

# YOUR FUTURE IN AI

*"The choice is yours."*
– The Matrix

This is the final chapter, and it's time for you to start thinking about your future in AI based on everything we've covered in this book. Remember, it's never too early or too late to plan for your career.

AI is everywhere, right? If you think AI is only relevant for programming and coding jobs, think again. It's being utilized in various roles, including marketing, sales, and design, whether that's fashion design, app design, or product design.

Analysts are also leveraging AI. Consider the financial analysts on Wall Street; they are increasingly using AI to select stocks, research companies, and summarize earnings calls from CEOs around the world to determine which companies are performing well or poorly. Their analysis heavily relies on AI as a tool. Even in the legal field, ethical AI experts use technology to help create regulations and build laws.

You don't need to limit yourself to technical or programming roles to engage with AI. It exists in every industry, from healthcare to gaming to law. As you contemplate your future career, think about how you can leverage AI for effective learning to better prepare yourself.

## Skills You Can Start Building Today

Some key skills you can start developing now, with the help of AI, include problem-solving. Are you someone who looks for solutions when faced with a challenge, or do you simply complain without taking action? Begin building that solution-oriented mindset.

Other important skills are critical thinking and communication. To enhance your communication skills, try recording yourself while speaking. Hearing your voice can be helpful in understanding how your message comes across differently than you perceive it in your mind.

A mindset shift is crucial. While you're likely already using AI in various applications, from smartphone apps to smart home devices to navigation systems, don't just use AI; understand how it works. Learn how you can utilize it to better prepare for your career.

## Future-Proofing Yourself

Finally, consider "future-proofing" yourself. Job roles will continually evolve. This change isn't solely due to AI; it has been happening throughout human history.

Consider the changes that occurred a few hundred years ago during the Industrial Revolution. Many traditional jobs disappeared, but at the same time, new jobs emerged. Similarly, the internet boom in the 1990s rendered many older, manual jobs obsolete as functions shifted online rather than remaining in the physical world.

Currently, we are experiencing a similar transformation with the rise of AI. Some jobs will inevitably vanish, while new ones will arise, which means you should prepare yourself to adapt to a changing environment rather than adhere to rigid rules that confine your daily life. If those rules shift, your entire life can be turned upside down.

Instead of focusing on maintaining a specific role, consider how you can leverage AI to solve problems, thereby creating new roles. For instance, if you build a product that addresses real-world issues and positively impacts lives using AI, you may create entirely new job descriptions to support that innovation.

Simply put, the future will belong to those who learn, adapt, and build with AI.

## Ways to Learn About AI

Learning doesn't have to be boring! Many students hear the word "learning" and think, *Forget it, that sounds dull.* However, there are exciting ways to learn.

Numerous YouTubers provide engaging explanations about how to use AI, and you can easily find and subscribe to their channels. There are also podcasts discussing various AI trends across different industries, as well as blogs and newsletters that you can subscribe to.

If you have a desire to learn, you can explore these resources in a way that suits your style. Enjoy watching videos? Search for them. Prefer listening? Subscribe to some podcasts. If you love reading, seek books, blogs, newsletters, and more.

When starting to learn about AI, remember you don't have to learn coding. While understanding the basics can be beneficial, it's not a requirement. There are plenty of no-code AI tools available that can help you accomplish your goals.

However, if you're interested in learning some fundamental coding, there are user-friendly platforms designed for beginners. AI playgrounds are also available, allowing you to experiment and test various ideas without any coding required.

Ultimately, as you explore different learning channels, focus on grasping the underlying concepts rather than memorizing information.

## Stay Curious and Keep Experimenting

Ask questions and stay curious. *Why is it this way? Why does it work like this?*

Inquire about everything: *Why this? Why that?*

The more questions you ask, the better.

Once you're curious, take the next step and experiment. If you're wondering how something works or want to try something new, go ahead and experiment!

Build it yourself using AI. Don't hesitate to involve others and see how they use it as well. Observe the results and learn from the experience.

I recommend a gradual approach. Challenge yourself to learn at least one new AI concept every week. Over time, this will accumulate into a significant body of knowledge.

**Key Takeaway:**

You don't need to know everything. You shouldn't stress about reaching the end goal all at once. Start small and keep learning consistently, and you'll get there.

**Key Takeaway for Part 4:**

Remember that AI is not just a tool; it comes with responsibilities. For many of you reading this book, it represents both a tool and a responsibility, and it can also be a remarkable opportunity.

**You can either watch the future unfold or take an active role in building it. The decision is yours.**

CONCLUSION

# NOW IT'S YOUR MOVE

When I started writing *The AI Survival Guide for Students*, my goal wasn't just to explain what AI is. There are plenty of resources that do that.

My intention was simple: **to help you understand AI well enough that it doesn't quietly control your decisions, your thinking, or your future.**

Because here's the truth: AI is not just another subject you'll study. It's a force that will shape:

- How you learn
- How you work
- How you communicate
- And even how you see the world

And the difference between being *impacted by AI* and being *in control of it* comes down to one thing:

**Understanding it.**

## What This Book Wanted You to See

Across these chapters, you explored AI from every angle, not just how it works but how it shows up in your life.

You saw that:

- AI is already everywhere, from your phone to your feed to your future.
- It learns from data, which means it can reflect both intelligence *and* bias.
- It doesn't "think" like humans; it predicts based on patterns.
- It can do incredible good at scale, but also cause harm if used irresponsibly.

You also saw something even more important:

**AI is powerful, but it is not perfect.**

## The Balance You Need to Remember

AI can:

- Help you learn faster
- Solve problems at scale
- Protect people and improve lives

But it can also:

- Mislead
- Manipulate
- Amplify unfairness
- Replace skills if you rely on it blindly

Which means the real skill isn't just *using AI.*

It's **using it wisely.**

## The Mindset That Sets You Apart

As you move forward, remember:

- Use AI to **understand**, not just to finish tasks.
- Question what you see, even if it looks real.
- Be intentional about what you click, share, and create.
- Don't outsource your thinking; enhance it.

Because the students who succeed in this new world won't be the ones who avoid AI.

They'll be the ones who **understand it better than everyone else.**

## Your Role in What Comes Next

You are not just users of AI.

You are:

- Future builders
- Decision-makers
- Problem-solvers
- And the generation that will decide what AI becomes

The systems being built today will shape tomorrow, but the people who understand those systems will shape *how* they're used.

## One Final Thought

There are two ways to live in the age of AI:

1. Let it guide you without realizing it
2. Understand it, and stay in control

This book was written to help you choose the second.

The future of AI is still being written.

**And whether you realize it or not, you're now part of that story.**

APPENDIX

# AI DOS & DON'TS: YOUR QUICK SURVIVAL GUIDE

**Use AI. Don't let it use you.**

**THE DOS (What You Should Do)**

**For Learning & School**

- **Use AI to understand concepts**, not just get answers.
- Ask AI to:
  - Explain step-by-step
  - Simplify complex topics
  - Give examples
- **Use AI for brainstorming**, not final submissions.
- Double-check AI answers with:
  - Textbooks
  - Teachers
  - Trusted sources

**For Thinking & Decision-Making**

- **Question what AI tells you.**

- Ask:
  - "Does this make sense?"
  - "Is this accurate?"
- Cross-verify important information.

**For Daily Use**

- Be aware of:
  - What you click
  - What you watch
  - What you share
- **Understand that algorithms learn from you.**
- Curate your feed intentionally.

**For Creativity**

- Use AI to:
  - Generate ideas
  - Explore creativity
  - Enhance your work
- Add your **own originality and voice.**

**For Privacy & Safety**

- Be careful what you:
  - Share
  - Upload
  - Type into AI tools
- Assume your data:
  - May be stored
  - May be used to improve systems

**For Responsibility**

- Think before creating or sharing content.
- Respect others':
  - Privacy
  - Identity
  - Consent
- Use AI in ways that:
  - Help people
  - Solve real problems

## THE DON'TS (What You Should Avoid)

**For School & Learning**

- Don't copy-paste AI answers as your own work.
- Don't rely on AI to do all your thinking.
- Don't use AI during exams if it is not allowed.

If you skip learning now, it catches up later.

**For Critical Thinking**

- Don't assume AI is always correct.
- Don't trust confident-sounding answers blindly.
- Don't rely on a single AI response for important decisions.

**For Social Media & Content**

- Don't believe everything you see online.
- Don't spread unverified or AI-generated content.

- Don't fall into algorithm "echo chambers."

**For AI-Generated Content**

- Don't create:
    - Fake images of people
    - Fake videos
    - Fake voices without consent
- Don't use AI to mislead or manipulate others.

**For Privacy**

- Don't share:
    - Personal data
    - Sensitive information
    - Passwords or private details
- Don't assume AI tools are completely private.

**For Ethics & Responsibility**

- Don't use AI to:
    - Cheat
    - Scam
    - Harm others
- Don't ignore the real-world impact of what you create.

## THE GOLDEN RULES (Remember These Always)

- AI is a tool, not a replacement for your thinking.
- Trust, but verify.

- Use AI to learn, not to shortcut.
- Just because you can, doesn't mean you should.
- The smarter you use AI, the more powerful you become.
- **Use it, don't let it use you.**

"If you don't learn how to think, AI will do it for you."

# THANK YOU FOR READING MY BOOK!

Just to say thanks for buying and reading my book, I would like to give you a few free bonus gifts, no strings attached!

**Scan the QR Code:**

TQRCG

*I appreciate your interest in my book and value your feedback, as it helps me improve future versions. I would appreciate it if you could leave your invaluable review on Amazon.com with your feedback. Thank you!*

www.ingramcontent.com/pod-product-compliance
Lightning Source LLC
LaVergne TN
LVHW100324210926
848480LV00003B/1247